The Carrier Bag Theory of Fiction

URSULA K. LE GUIN

Published in this revised single edition in 2024
by cosmogenesis
cosmogenesis.shop

'The Carrier Bag Theory of Fiction' by Ursula K. Le Guin was first
published in *Women of Vision*, 1988

ISBN-13: 9781838003982

Design by Cecilia Serafini
Revised cover design by Kathryn Politis
Typeset in Adobe Caslon Pro by Cecilia Serafini
Printed in China by Everbest Printing Co. Ltd

This edition and selection first published by Ignota (2018-2024), a press created by Sarah Shin and Ben Vickers. An archive of Ignota's works can be found at ignota.org.

3 5 7 9 10 8 6 4 2

Contents

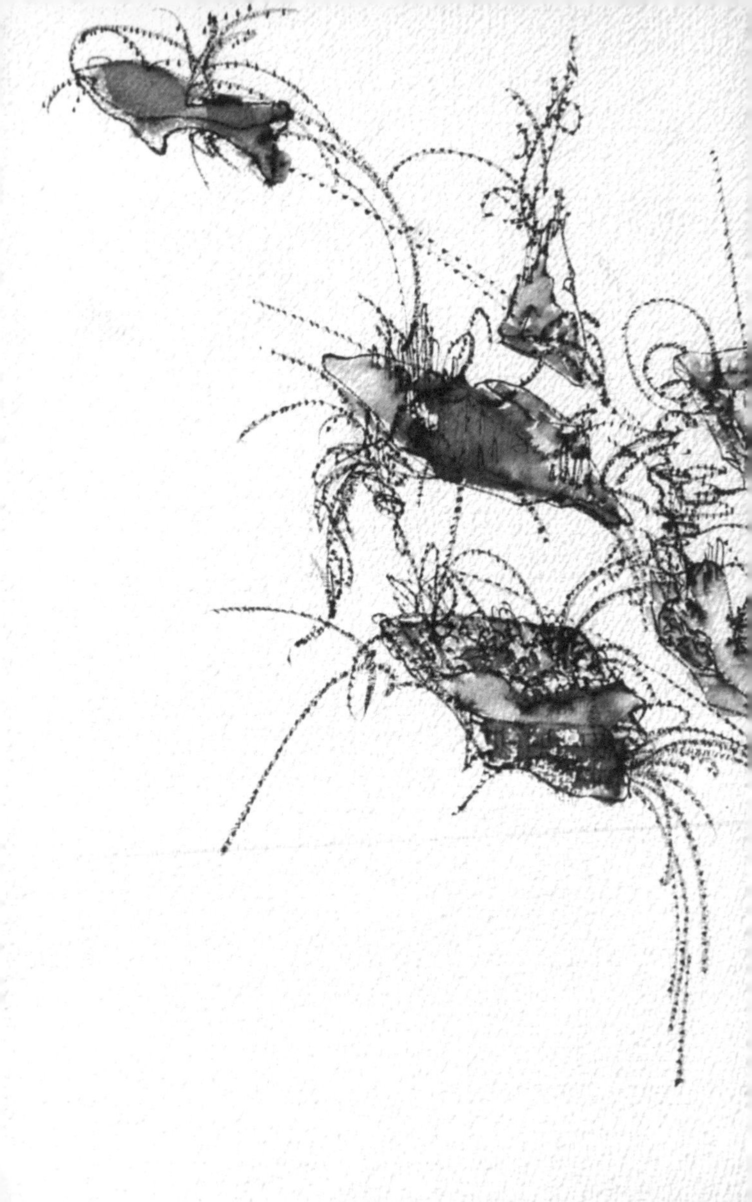

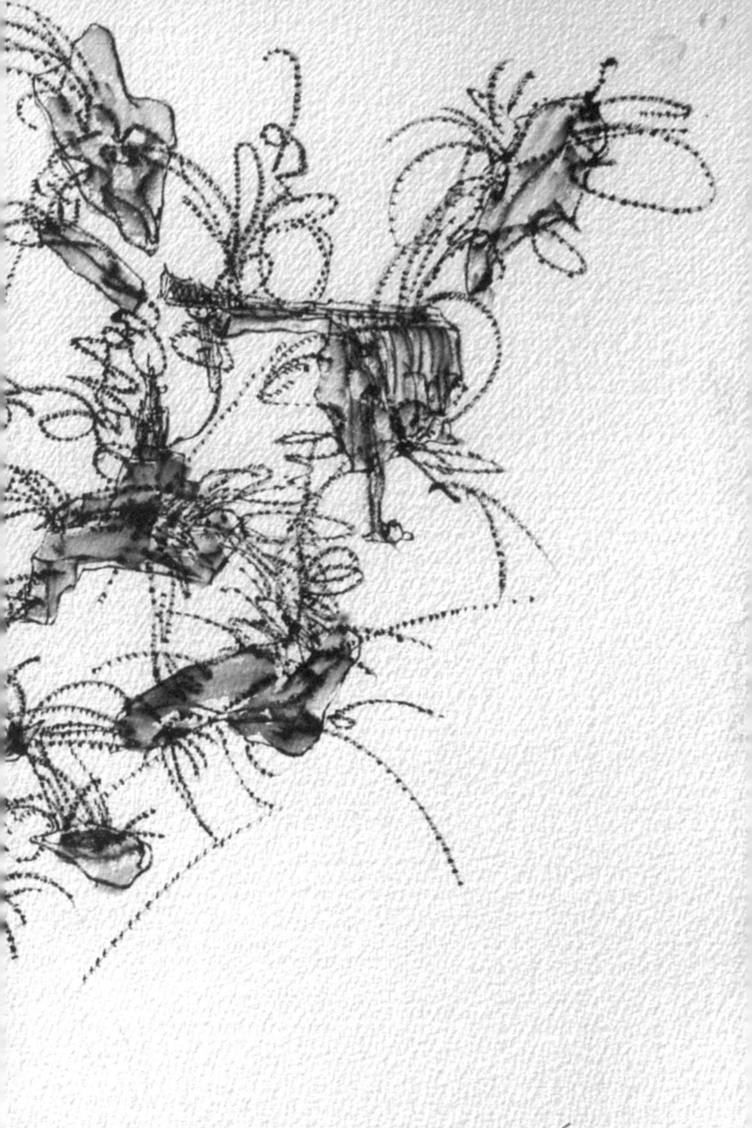

Introduction

Receiving Three Mochilas in Colombia
Carrier Bags for Staying with the Trouble Together

BY DONNA HARAWAY
IN HONOUR OF URSULA K. LE GUIN

Ursula K. Le Guin's *The Carrier Bag Theory of Fiction* touched me at my core when I first read it in the late 1980s[1] – and it still undoes and redoes me in the knotting of life stories for going on together in times of immense danger. In August 2019, I came home from an intense, two-week working trip to Bogotá, Bucaramanga, and Santa Marta. With warm

1. Le Guin's 'Carrier Bag Theory of Fiction' shaped my thinking about narrative in evolutionary theory and of the figure of the woman gatherer in my book about the history of primate behavior studies, *Primate Visions: Gender, Race and Nature in the World of Modern Science*, New York: Routledge, 1989. Le Guin learned about the Carrier Bag Theory of human evolution from Elizabeth Fisher in that period of large, brave, speculative, worldly stories that burned in feminist theory in the 1970s and 1980s (Elizabeth Fisher, *Women's Creation*, New York: McGraw-Hill, 1975).

Like speculative fabulation, speculative feminism was and is an SF practice. All quotations from Le Guin's story are from Ursula K. Le Guin, *The Carrier Bag Theory of Fiction*, London: Ignota, 2019.

generosity, my friends and colleagues in anthropology, science and technology studies, the arts and environmental humanities engaged me in their work with and for: land and water protection and recovery; Indigenous flourishing; non-violent space for urban street people; compensation for *campesinos* ruined by crop destruction, drug wars and development projects; supporting Afro-Colombian lives and cultures, and deep and broad natural social justice; and care in a country of such layered complexity and resurgent pain. My friends showed me ordinary people working and playing to stitch, knot, weave and embroider both intimate and public peace, even as it unravels yet again. My Colombian friends understand in their bones the importance of what Le Guin taught us to do, to keep telling the 'other story, the untold one, the life story'.

It matters what stories we tell to tell other stories with; it matters what concepts we think to think other concepts with. It matters wherehow ouroboros swallows its tale, again. That's how worlding gets on with itself in dragon time. A brave student of dragons, Le Guin's stories are capacious bags for collecting, carrying, and telling the stuff of living. 'A leaf a gourd a shell a net a bag a sling a sack a bottle a pot a box a container. A holder. A recipient.'

Offering such reciprocal stories, my companions gave me three very different carrier bags, three *mochilas*, for collecting the particular, powerful things needed to nurture becoming-with each other for ongoing processes of living and dying otherwise. The stories these bags gather can mute the killing stories, the life-sucking weaponised sciences and technologies, the drug-and-money-soaked vampire markets, and all the other prick tales that labour to make history in their own image. None of these bags is a utopia outside the killing fields; quite the opposite. They each situate those who make and those who carry the *mochila* in worlds that are at stake now. The *mochilas* strengthen the people who make and use them. These carrier bags make their people more worldly, more able to discern and tell what is really happening and how it can still be different. Each *mochila* grows from, and demands a response to, the urgent questions about how to tell stories that can help remake history for the kinds of living and dying that deserve thick presents and rich futures. In the spirit of Le Guin's insight that the fitting shape of a story is a sack, a hollowed-out container to hold things that bear meanings and enable relationships, each *mochila* is a bag for the gripping tales and strange realism – the serious fiction, the science fiction, the SF – required for inhabiting the worlds of seeds and stars. These are the worlds full of human and more-than-human people of many kinds, and species with better things to do than kill each other while sucking their planet dry.

Let me begin to tell stories that each *mochila*, each carrier bag, made palpable for me. To carry, to wear, any of these bags is to enter into the knotting of capacities to respond, to become-with each other in the untold stories we need.

FIGURE 1

1. 'Flore-Ser', literally 'flower-be', which sounds like 'florecer', to flourish, is embroidered on the blue linen of a bag made by the women of AMARÚ, la Asociación de Mujeres Defensoras del Agua y la Vida (the Association of Women in Defence of Water and Life).

2. The bag with wide stripes in naturally coloured sheep's wool was intricately knotted by an elder Arhuaco woman in the Sierra Nevada de Santa Marta. Her granddaughter Ati, president of the Indigenous Students Association of La Universidad de Magdalena in Santa Marta, sold this bag to the ethnographers William Martinez Dueñas and Astrid Lorena Perafan Ledezma, who helped bring me to Colombia and who gave the *mochila* to me.

Figure 1

3. The rust-coloured, geometrically patterned *mochila* was crocheted with cotton by Wayúu women in La Guajira in northeastern Colombia.

I'll call my first bag 'Flore-Ser'; she already bears this promising name for flourishing in ambiguously subjunctive and imperative verb form embroidered on her front. My friend and colleague Tania Pérez-Bustos gave me this bag minutes after she picked me up at the airport in Bogotá in August 2019. Flore-Ser's habitat is textile activism by a collective that works for environmental justice and sexual and reproductive rights. At first, I saw only the delicate white flower inside the teal-coloured uterus. Over two weeks I learned to see and feel what Tania, an ethnographer and science and technology studies scholar, taught me about the gestation of oddkin. I collected the pieces of my visit in this bag so that I might begin to learn to join with the people of Colombia in their work and play for multispecies environmental and social justice and care. In her work, Tania brings together fibre arts and their people, in craft design and in engineering design. She especially studies the 'entangled practices of unraveling and mending' – of lives and of fabric – in calado embroidery in Cartago, Colombia.[2] Tania relates how the women she works with show her how slow hand stitching, in these hard times, is vital for personal and intimate healing, for rebinding destroyed communities and for telling the histories of land, water, displacement and still possible futures.

Flore-Ser is made by the women of El Movimiento Rios Vivos Antioquia as a means of both subsistence and resistance in the face of the immense hydroelectric project, Hidroituango, being constructed in the Cauca River basin on their lands and waters. The women both come from, and ally with, the diverse human people and the vast variety of other kinds of living beings in one of the last well-conserved, dry tropical forest naturalcultural ecosystems of Latin America. Facing a wave of increasing violence against those who are defending their territory and water by opposing the project, these women defend what they call their *cuerpo-territoria* (body-land) in the fight against what they identify as masculine domination in all its forms. Storytelling in embroidery

2. Tania Pérez-Bustos, 'Thinking with Care: Unraveling and Mending in an Ethnography of Craft Embroidery and Technology', *Revue d'Anthropologie des Connaissances* 11:1, 2017.

is not a luxury; it is making room for the life story in what Le Guin called 'the bag of stars'. They collected me and my companions in their bag. Failure to respond is not a choice if I wear that bag. I brought this bag home with me to Santa Cruz, California, where we have much to do to embroider stories together effectively.

My second bag, with wide stripes in naturally coloured sheep's wool, was intricately knotted by an elder Arhuaco woman in the Sierra Nevada de Santa Marta. Each design depicts a crucial story taught by one generation of women to the next. Her granddaughter Ati, president of the Indigenous Students Association of La Universidad de Magdalena in Santa Marta, sold this bag to the ethnographers William Martinez Dueñas and Astrid Lorena Perafan Ledezma, who helped bring me to Colombia and who gave the *mochila* to me. Ati participated in the public conversation with me and other colleagues at the university. Today these bags relate closely to young Arhuaco people claiming their identities in alliance with other Indigenous people (Kogui, Wiwa and Kankuamo) for control of their land and stories in the face of ongoing settlement and development schemes by outsiders, imposed illicit crops and government fumigation, assassinations and intimidation, climate change, ecotourism, dam projects and mining. There is no easy way to inhabit the stories this bag tells, but it attaches me to audacious, ongoing knotting for lives and land. Wearing the bag means learning to listen and respond to the life story.

My third bag collects similar narratives, but like all stories, these are situated someplace in particular, not everywhere or nowhere in killing abstraction. The rust-coloured, geometrically patterned *mochila* was crocheted with cotton by a particular Wayúu woman in La Guajira in northeastern Colombia. I do not know her name. The bag was given to me by the Colombian anthropologist Leonardo Montenegro, who is working to support Wayúu communities in the struggle against the multinationals Anglo-American, Glencore and BHP Billeton for land, water and survival. The situation is drastic; children and animals die for lack of food and water. Crops fail in endless drought. The poverty and thirst are not a natural fact; they are the fruit of the second largest opencast coal mine in the world, the Cerréjon coal mine, with its endless lust for water for processing and transport of this earth-roasting fuel. For example, the damming of the Rancheria river enables Cerrejón to use 17 million litres of water a day while each resident of La Guajira is left with an average of 0.7 litres per day to live on.[3] Threats of death from paramilitaries against those opposing the mine are common in the attempt to clear the Wayúu off their ancestral land. But the African-descendent and Indigenous communities continue to join together for their lives and land, and they have built some strong alliances in the world. I too know a little bit of their story now. I have been collected in this carrier bag. The burning question is how to join in telling the needed stories, building the needed worlds and muting the deadly ones. The lush feel of the soft fabric, with its burnt umber colors and designs, offers life-sustaining corporal sensual sustenance that is needed in making the untold stories strong.

I close with a provocative multi-authored study among the Agta, a group of modern hunter-gatherers in the Philippines, who bring us back to Le Guin's willingness to entertain big evolutionary stories in her 'Carrier Bag Theory of Fiction'.[4] The researchers identified and listened to

3. The Gaia Foundation, 'Indigenous Colombians Threatened with Death for Defending Water, Territory, and Life', gaiafoundation.org (accessed 15 September 2019).

4. Daniel Smith, Philip Schlaepfer, Katie Major, Mark Dyble, Abigail E. Page, James Thompson, Nikhil Chaudhary, Gul Deniz Salali, Ruth Mace, Leonora Astete, Marilyn Ngales, Lucia Vincius and Andrea Ramberg Migliano, 'Cooperation and the Evolution of Hunter-Gatherer Storytelling', *Nature Communications* 8:1853, 2018.

many stories and storytellers over a long period of time and designed typical social scientific instruments to find out what people in these communities valued most. It turned out that the Agta responders value storytellers over all other kinds of people, no matter how useful and functional other sorts of persons might be in their society. The authors insisted on a biological fitness explanation for this finding; good storytellers turn out to be more attractive mates and people like to give useful things to skilled storytellers. But without much sympathy for the endless domination by the tale of competitive biological reproductive advantage for anything that looks like cooperation, I still find much to love in this study. The authors report that camps with a greater proportion of adept storytellers showed increased levels of cooperation. People prefer to live in camps with good storytellers. Furthermore, the great majority of the stories emphasise cooperation and sexual and social equality. Adults and children listen to and want to hear these stories; they are full of exciting actors and rich happenings. Food is important, of course; but people say that they like good storytellers even more than good foragers, and their behaviour seems to confirm their self evaluation. The stories appear to be valued for the ways they increase empathy and more welcoming perspectives towards others, even strangers. The authors speculate that storytelling might well be fundamental to organising and promoting cooperation in human evolution.

I find myself quickly falling down the rabbit hole of just-so stories that I prefer. But at the least, the study of Agta relations to their storytellers is provocative. It cries out for more study, if the people are interested in the research, participate meaningfully in its design and dissemination, and find the whole process conducive to their wellbeing. With firm resolve, I resist universalising and reifying the tale of the story-loving Agta. But with Le Guin and Elizabeth Fisher, and ongoing generations of storytellers intent on singing of the powerful practices of living and dying well with each other in tales that have the shape of a well-wrought urn or cleverly knotted bag, I am heartened by the Agta. Their love of storytelling collects possibilities of recomposing lives and making new sorts of kin in hard times. The Storied Ones are powerful transformers and inventors of patterns for still possible flourishing. Flore-Ser.

As Le Guin told us in her mighty little essay, 'It sometimes seems that that [heroic] story is approaching its end. Lest there be no more telling of stories at all, some of us out here in the wild oats, amid the alien corn, think we'd better start telling another one, which maybe people can go on with when the old one's finished. Maybe. The trouble is, we've all let ourselves become part of the killer story, and so we may get finished along with it. Hence it is with a certain feeling of urgency that I seek the nature, subject, words of the other story, the untold one, the life story.'

BIBLIOGRAPHY

Fisher, Elizabeth, *Women's Creation*, New York: McGraw-Hill, 1975.

Haraway, Donna, *Primate Visions: Gender, Race and Nature in the World of Modern Science*, New York: Routledge, 1989.

Le Guin, Ursula K., *The Carrier Bag Theory of Fiction*, London: Ignota, 2019.

Pérez-Bustos, Tania, 'Thinking with Care: Unraveling and Mending in an Ethnography of Craft Embroidery and Technology', *Revue d'Anthropologie des Connaissances* 11:1, 2017.

Smith, Daniel, Schlaepfer, Philip, Major, Katie, Dyble, Mark, Page, Abigail E., Thompson, James, Chaudhary, Nikhil, Deniz Salali, Gul, Mace, Ruth, Astete, Leonora, Ngales, Marilyn, Vincius, Lucia, and Ramberg Migliano, Andrea, 'Cooperation and the Evolution of Hunter-Gatherer Storytelling', *Nature Communications* 8: 1853, 2018.

The Carrier Bag Theory of Fiction

URSULA K. LE GUIN

In the temperate and tropical regions where it appears that hominids evolved into human beings, the principal food of the species was vegetable. Sixty-five to eighty percent of what human beings ate in those regions in Paleolithic, Neolithic and prehistoric times was gathered; only in the extreme Arctic was meat the staple food. The mammoth hunters spectacularly occupy the cave wall and the mind, but what we actually did to stay alive and fat was gather seeds, roots, sprouts, shoots, leaves, nuts, berries, fruits and grains, adding bugs and mollusks and netting or snaring birds, fish, rats, rabbits and other tuskless small fry to up the protein. And we didn't even work hard at it – much less hard than peasants slaving in somebody else's field after agriculture was invented, much less hard than paid workers since civilisation was invented. The average prehistoric person could make a nice living in about a fifteen-hour work week.

Fifteen hours a week for subsistence leaves a lot of time for other things. So much time that maybe the restless ones who didn't have a baby around to enliven their life, or skill in making or cooking or singing, or very interesting thoughts to think, decided to slope off and hunt mammoths. The skillful hunters then would come staggering back with a load of meat, a lot of ivory and a story. It wasn't the meat that made the difference. It was the story.

It is hard to tell a really gripping tale of how I wrested a wild-oat seed from its husk, and then another, and then another, and then another, and then another, and then I scratched my gnat bites, and Ool said something funny, and we went to the creek and got a drink and watched newts for a while, and then I found another patch of oats ... No, it does not compare, it cannot compete with how I thrust my spear deep into the titanic hairy flank while Oob, impaled on one huge sweeping tusk, writhed screaming, and blood spouted everywhere in crimson torrents, and Boob was crushed to jelly when the mammoth fell on him as I shot my unerring arrow straight through eye to brain.

That story not only has Action, it has a Hero. Heroes are powerful. Before you know it, the men and women in the wild-oat patch and their kids and the skills of the makers and the thoughts of the thoughtful and the songs of the singers are all part of it, have all been pressed into service in the tale of the Hero. But it isn't their story. It's his.

When she was planning the book that ended up as *Three Guineas*, Virginia Woolf wrote a heading in her notebook, 'Glossary'; she had thought of reinventing English according to a new plan, in order to tell a different story. One of the entries in this glossary is *heroism*, defined as 'botulism'. And *hero*, in Woolf's dictionary, is 'bottle'. The hero as bottle, a stringent re-evaluation. I now propose the bottle as hero.

Not just the bottle of gin or wine, but bottle in its older sense of container in general, a thing that holds something else.

If you haven't got something to put it in, food will escape you – even something as uncombative and unresourceful as an oat. You put as many as you can into your stomach while they are handy, that being the primary container; but what about tomorrow morning when you wake up and it's cold and raining and wouldn't it be good to have just a few handfuls of oats to chew on and give little Oom to make her shut up, but how do you get more than one stomachful and one handful home? So you get up and go to the damned soggy oat patch in the rain, and wouldn't it be a good thing if you had something to put Baby Oo Oo in so that you could pick the oats with both hands? A leaf a gourd a shell a net a bag a sling a sack a bottle a pot a box a container. A holder. A recipient.

The first cultural device was probably a recipient . . . Many theorisers feel that the earliest cultural inventions must have been a container to hold gathered products and some kind of sling or net carrier.

So says Elizabeth Fisher in *Women's Creation* (McGraw-Hill, 1975). But no, this cannot be. Where is that wonderful, big, long, hard thing, a bone, I believe, that the Ape Man first bashed somebody with in the movie and then, grunting with ecstasy at having achieved the first proper murder, flung up into the sky, and whirling there it became a space ship thrusting its way into the cosmos to fertilise it and produce at the end of the movie a lovely fetus, a boy of course, drifting around the Milky Way without (oddly enough) any womb, any matrix at all? I don't know. I don't even care. I'm not telling that story. We've heard it, we've all heard all about all the sticks and spears and swords, the things to bash and poke and hit with, the long, hard things, but we have not heard about the thing to put things in, the container for the thing contained. That is a new story. That is news.

And yet old. Before – once you think about it, surely long before – the weapon, a late, luxurious, superfluous tool; long before the useful knife and axe; right along with the indispensable whacker, grinder and digger – for what's the use of digging up a lot of potatoes if you have nothing to lug ones you can't eat home in – with or before the tool that forces energy outward, we made the tool that brings energy home. It makes sense to me. I am an adherent of what Fisher calls the Carrier Bag Theory of human evolution.

This theory not only explains large areas of theoretical obscurity and avoids large areas of theoretical nonsense (inhabited largely by tigers, foxes and other highly territorial mammals); it also grounds me, personally, in human culture in a way I never felt grounded before. So long as culture was explained as originating from and elaborating upon the use of long, hard objects for sticking, bashing and killing, I never thought that I had, or wanted, any particular share in it. ('What Freud mistook for her lack of civilisation is woman's lack of *loyalty* to civilisation', Lillian Smith observed.) The society, the civilisation they were talking about, these theoreticians, was evidently theirs; they owned it, they liked it; they were human, fully human, bashing, sticking, thrusting, killing. Wanting to be human too, I sought for evidence that I was;

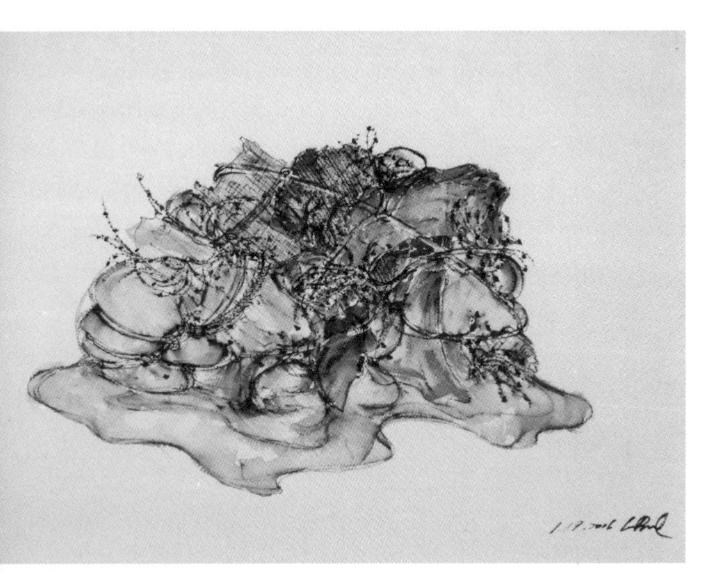

but if that's what it took, to make a weapon and kill with it, then evidently I was either extremely defective as a human being, or not human at all.

That's right, they said. What you are is a woman. Possibly not human at all, certainly defective. Now be quiet while we go on telling the Story of the Ascent of Man the Hero.

Go on, say I, wandering off towards the wild oats, with Oo Oo in the sling and little Oom carrying the basket. You just go on telling how the mammoth fell on Boob and how Cain fell on Abel and how the bomb fell on Nagasaki and how the burning jelly fell on the villagers and how the missiles will fall on the Evil Empire, and all the other steps in the Ascent of Man.

If it is a human thing to do to put something you want, because it's useful, edible or beautiful, into a bag, or a basket, or a bit of rolled bark or leaf, or a net woven of your own hair, or what have you, and then take it home with you, home being another, larger kind of pouch or bag, a container for people, and then later on you take it out and eat it or share it or store it up for winter in a solider container or put it in the medicine bundle or the shrine or the museum, the holy place, the area that contains what is sacred, and then next day you probably do much the same again – if to do that is human, if that's what it takes, then I am a human being after all. Fully, freely, gladly, for the first time.

Not, let it be said at once, an unaggressive or uncombative human being. I am an aging, angry woman laying mightily about me with my handbag, fighting hoodlums off. However I don't, nor does anybody else, consider myself heroic for doing so. It's just one of those damned things you have to do in order to be able to go on gathering wild oats and telling stories.

It is the story that makes the difference. It is the story that hid my humanity from me, the story the mammoth hunters told about bashing, thrusting, raping, killing, about the Hero. The wonderful, poisonous story of Botulism. The killer story.

It sometimes seems that that story is approaching its end. Lest there be no more telling of stories at all, some of us out here in the wild oats, amid the alien corn, think we'd better start telling another one, which maybe people can go on with when the old one's finished. Maybe. The trouble is, we've all let ourselves become part of the killer story, and so we may get finished along with it. Hence it is with a certain feeling of urgency that I seek the nature, subject, words of the other story, the untold one, the life story.

It's unfamiliar, it doesn't come easily, thoughtlessly to the lips as the killer story does; but still, 'untold' was an exaggeration. People have been telling the life story for ages, in all sorts of words and ways. Myths of creation and transformation, trickster stories, folktales, jokes, novels...

The novel is a fundamentally unheroic kind of story. Of course the Hero has frequently taken it over, that being his imperial nature and uncontrollable impulse, to take everything over and run it while making stern decrees and laws to control his uncontrollable impulse to kill it. So the Hero has decreed through his mouthpieces the Lawgivers, first, that the proper shape of the narrative is that of the arrow or spear, starting *here* and going straight *there* and THOK! hitting its mark (which drops dead); second, that the central concern of narrative, including the novel, is conflict; and third, that the story isn't any good if he isn't in it.

I differ with all of this. I would go so far as to say that the natural, proper, fitting shape of the novel might be that of a sack, a bag. A book holds words. Words hold things. They bear meanings. A novel is a medicine bundle, holding things in a particular, powerful relation to one another and to us.

One relationship among elements in the novel may well be that of conflict, but the reduction of narrative to conflict is absurd. (I have read a how-to-write manual that said, 'A story should be seen as a battle', and went on about strategies, attacks, victory, etc.) Conflict, competition, stress, struggle, etc., within the narrative conceived as carrier bag/belly/box/house/medicine bundle, may be seen as necessary elements of a whole which itself cannot be characterised either as conflict or as harmony, since its purpose is neither resolution nor stasis but continuing process.

Finally, it's clear that the Hero does not look well in this bag. He needs a stage or a pedestal or a pinnacle. You put him in a bag and he looks like a rabbit, like a potato.

That is why I like novels: instead of heroes they have people in them.

So, when I came to write science-fiction novels, I came lugging this great heavy sack of stuff, my carrier bag full of wimps and klutzes, and tiny grains of things smaller than a mustard seed and intricately woven nets which when laboriously unknotted are seen to contain one blue pebble, an imperturbably functioning chronometer telling the time on another world and a mouse's skull; full of beginnings without ends, of initiations, of losses, of transformations and translations, and far more tricks than conflicts, far fewer triumphs than snares and delusions; full of space ships that get stuck, missions that fail and people who don't understand. I said it was hard to make a gripping tale of how we wrested the wild oats from their husks, I didn't say it was impossible. Who ever said writing a novel was easy?

If science fiction is the mythology of modern technology, then its myth is tragic. 'Technology', or 'modern science' (using the words as they are usually used, in an unexamined shorthand standing for the 'hard' sciences and high technology founded upon continuous economic growth), is a heroic undertaking, Herculean, Promethean, conceived as triumph, hence ultimately as tragedy. The fiction embodying this myth will be, and has been, triumphant (Man conquers earth, space, aliens, death, the future, etc.) and tragic (apocalypse, holocaust, then or now).

If, however, one avoids the linear, progressive, Time's-(killing)-arrow mode of the Techno-Heroic, and redefines technology and science as primarily cultural carrier bag rather than weapon of domination, one pleasant side effect is that science fiction can be seen as a far less rigid, narrow field, not necessarily Promethean or apocalyptic at all, and in fact less a mythological genre than a realistic one.

It is a strange realism, but it is a strange reality.

Science fiction properly conceived, like all serious fiction, however funny, is a way of trying to describe what is in fact going on, what people actually do and feel, how people relate to everything else in this vast sack, this belly of the universe, this womb of things to be and tomb of things that were, this unending story. In it, as in all fiction, there is room enough to keep even Man where he belongs, in his place in the scheme of things; there is time enough to gather plenty of wild oats and sow them too, and sing to little Oom, and listen to Ool's joke, and watch newts, and still the story isn't over. Still there are seeds to be gathered, and room in the bag of stars.

ILLUSTRATIONS BY LEE BUL

p. 4-5: Drawing for *Mon Grand Récit (abstract crystal sculpture 1)*, 2006

p. 24: Study for *Mon Grand Récit (Circular Prison)*, 2006

p. 29: Study for *Mon Grand Récit (Fondation Cartier)*, 2006

p. 36-37: Drawing for *After Bruno Taut (crystal architectural landscapes)*, 2006